Extraction of Magic Mushroom

A Detailed Guide in Exploring the Cultivation and Utilization of Magic Mushrooms

Maria Cater

Copyright@2024 Maria Cater all right reserved. No part of this publication may be reproduced in any form or means without prior written permission from the copy right holder

- Chapter One 5
- Chapter Two 10
  - Magic mushroom taxonomy and categorization 10
- Chapter Three 20
  - The physiological processes behind psychedelic effects 20
- Chapter four 26
  - Methods of Cultivation 26
- Chapter Five 45
  - The growth-promoting environment 45
- Chapter six 51
  - Key psychoactive alkaloids (such as psilocybin and psilocin) 51
- Chapter Seven 58
  - These substances' effects on the human brain 58
- Chapter Eight 72
  - Techniques for Extracting 72
- Chapter Nine 79
  - Traditional techniques for extracting 79
- Chapter Ten 86
  - Extraction techniques employed in labs 86

Chapter eleven 93
  Purification and Isolation of Psilocybin 93
Chapter Twelve 106
  Examining spirituality and consciousness ..106

Chapter One
What magic mushrooms are and why they matter

Psilocybin mushrooms, or magic mushrooms as they are known, are a class of fungi that consist of psilocybin and other hallucinogenic substances. These substances are well-known for their hallucinogenic properties, which when used can cause changes in mood, perception, and awareness.

The lengthy history of using magic mushrooms for therapeutic, spiritual, and religious purposes throughout many cultures and civilizations accounts for their relevance. For thousands of years, indigenous peoples have used magic mushrooms in religious rites and healing rituals in areas like Mesoamerica. These mushrooms were revered and thought to improve spiritual experiences, encourage healing, and permit divine connection.

These days, magic mushrooms are becoming more and more well-known for their potential therapeutic uses in treating a range of mental health conditions, including as addiction, PTSD, depression, and anxiety. Promising findings from studies on the therapeutic benefits of psilocybin-assisted therapy have raised interest in the potential use of these fungi in the treatment of mental illness.

Furthermore, because of their distinct chemical makeup and mental consequences, magic mushrooms have aroused scientific interest. Researchers have learned more about consciousness, perception, and the brain's reaction to psychoactive substances through studies examining the neurobiological mechanisms underpinning the psychedelic experience.

However, because magic mushrooms are regarded as a restricted substance in many nations, there are additional ethical and legal issues to consider. Legal limitations still prevent access to and study on these fungi for medical and recreational uses, despite increased interest in and lobbying for their decriminalization and legalization.

In general, magic mushrooms are significant cultural and scientific phenomena that raises significant issues regarding the morality, lawfulness, and public perceptions of psychedelic drugs in addition to providing insights into human consciousness, spirituality, and mental health.

Chapter Two

Magic mushroom taxonomy and categorization

Magic mushroom taxonomy and classification refers to their systematic grouping according to different attributes like morphology, genetics, and biochemical makeup. This is a summary of the classification scheme for magic mushrooms:

1. Kingdom: Molds, yeasts, and mushrooms are among the organisms that make up the kingdom Fungi, which includes magic mushrooms. Different from bacteria, plants, and animals, fungi get their nourishment from the breakdown of organic materials or by symbiotically interacting with other living things.

2. Division: Magic mushrooms are members of the Basidiomycota division, also referred to as club fungi, within the kingdom Fungi. This group of fungi is made up of a wide variety of species, such as mushrooms, toadstools, and bracket fungi, which are distinguished by the existence of spore-producing reproductive organs called basidia.

3. Class: The class Agaricomycetes, which includes a

wide variety of fungi, including several mushroom species, is where magic mushrooms are further categorized. The development of basidiocarps, or fruiting bodies, which release spores on specialized structures known as basidia, is a defining characteristic of agaricomycetes.

4. Order: Magic mushrooms are usually placed in the order Agaricales within the class Agaricomycetes. This order contains a number of well-known mushroom species, including button mushrooms (Agaricus bisporus), as well as some hallucinogenic mushroom species.

The presence of gilled mushrooms with spore-bearing structures known as basidia characterizes the order Agaricales.

5. Family: Magic mushrooms fall into a number of families within the Agaricales order based on their unique traits and evolutionary connections. Magic mushrooms are commonly found in the Psilocybeaceae, Hymenogastraceae, and Strophariaceae groups.

6. Genus and Species: In conclusion, the physical characteristics, genetic relationships, and biochemical makeup of magic mushrooms enable their classification into distinct genera and species. Psilocybe cubensis and Psilocybe semilanceata are two examples of the numerous species of magic mushrooms in the genus Psilocybe that are well-known for their hallucinogenic qualities.

Mushroom anatomy and life cycle

Magic mushrooms are no exception when it comes to their anatomy and life cycle, which include multiple unique stages and structures. This is a synopsis:

1. Hyphae: Known as microscopic filaments, hyphae are the first stage of a mushroom's life cycle. The mycelium, a network made up of these hyphae, develops within or beneath its substrate. Mycelium is the fungus's feeding and reproductive structure, absorbing nutrients from the surrounding environment.

2. Fruiting Body: The visible portion of the mushroom is produced by the mycelium under the correct circumstances. The fruiting body normally consists of a cap (pileus) and a stalk (stipe), though species-to-species variations in form are common. The spore-producing gills, holes, or teeth are found inside the cap.

3. Spore Production: A process known as meiosis is used by microscopic cells called basidia inside the gills or other spore-bearing structures to manufacture spores. As the mushroom ages and its cap splits open, these spores are discharged into the surrounding environment. The main method by which mushrooms reproduce is by spores, which enable them to proliferate and occupy new areas.

4. Reproductive Cycle: Asexual and sexual reproduction is both a part of a mushroom's life cycle. When two suitable hyphae unite to form a dikaryotic mycelium a structure with two nuclei per cell—sexual reproduction takes place. The fruiting body, which is created when the nuclei in the basidia fuse, is derived from this mycelium. Asexual reproduction can also take place by producing specialized structures known as conidia or by breaking apart the mycelium.

5. Ecological Role: As decomposers that recycle nutrients back into the environment and break down organic debris, mushrooms are essential to ecosystems. They associate with plants in symbiotic ways, such as mycorrhizal connections, in which they trade nutrients with the plants they inhabit. In addition, mushrooms support the structure and health of the soil and provide food for a variety of creatures.

6. Variations in the Life Cycle: The life cycle of mushrooms can differ depending on the species and the surrounding circumstances. While some mushrooms are perennial and can live for several years, others are annuals and complete their life cycle in just one growing season. Furthermore, environmental elements that affect the timing and success of mushroom fruiting include temperature, moisture content, and nutrient availability.

It is crucial to comprehend the structure and life cycle of mushrooms in order to cultivate, identify, and preserve their natural habitat. It sheds light on the ecological roles that mushrooms play in the dynamics of ecosystems and the cycling of nutrients.

Chapter Three

The physiological processes behind psychedelic effects

Psilocybin and psilocin, among other substances, are mostly responsible for the psychedelic effects of magic mushrooms, which result from their interaction with the serotonin pathway in the brain. An outline of the biological mechanisms at play is provided below:

1. Serotonin Receptor Activation: The main hallucinogenic ingredient in magic mushrooms, psilocybin, shares structural similarities with serotonin, a neurotransmitter involved in mood regulation, perception of sensory information, and thought processes. The primary mechanism by which psilocybin and its metabolite, psilocin, work in the brain is through binding to and activating serotonin receptors, specifically the serotonin 2A receptor (5-HT2A). This activation ultimately results in altered perception and cognition through modifications to neuronal signaling and neurotransmitter release.

2. Neurotransmitter Modulation: Psilocybin and psilocin interact with dopamine, glutamate, and norepinephrine systems in the brain in addition to serotonin receptors. The complex effects of psychedelics on emotion, perception, and consciousness are partly attributed to these interactions. For instance, altered glutamate transmission may be responsible for alterations in sensory processing and perception, while elevated dopamine release in specific brain regions may add to the euphoric and rewarding effects of psychedelics.

3. Disruption of Default Mode Network (DMN): The ability of psychedelics to interfere with the functioning of the default mode network (DMN), a network of brain regions implicated in introspection, self-referential thought, and ego-related activities, is one of their most intriguing side effects. Studies using functional neuroimaging have demonstrated that psychedelics such as psilocybin cause a transient disruption of the DMN, which results in changes in consciousness and a collapse of the concept of ego. The dramatic shifts in perception, self-awareness, and mystical experiences that psychedelic users report are thought to be caused by this disturbance.

4. Neuroplasticity and Synaptic Remodeling: New study indicates that psychedelics may enhance synaptic remodeling and neuroplasticity in the brain, which could result in long-lasting alterations to behavior and neural connectivity. Research on animals has indicated that psychedelics, such as psilocybin, have the ability to promote the development of new neurons and synapses, especially in areas of the brain linked to memory, learning, and emotional control. These neuroplastic effects may have therapeutic implications for disorders defined by maladaptive thought processes and faulty brain circuits, such as PTSD, depression, and anxiety.

Magic mushrooms' overall psychedelic effects are caused by their interactions with several neurotransmitter systems, modification of brain network activity, and enhancement of neuroplasticity. Research in this area has the potential to clarify the therapeutic potential of these substances as well as their effects on consciousness and cognition, even though our knowledge of the molecular mechanisms behind psychedelics is still developing.

Chapter four

Methods of Cultivation

The process of growing magic mushrooms is establishing the right conditions for the fungus's mycelium, or vegetative portion, to grow, and then encouraging fruiting, which is when the mushrooms actually mature. An outline of cultivation methods is provided below:

1. Preparing the Substratum:

• Choose an appropriate substrate: Supplemented sawdust, straw, or sterilized grain (such as wheat, rye, or millet) are common substrates.

- Sterilize the substrate: Generally, pressure cooking or steam sterilization are used to sterilize the substrate in order to stop it from becoming contaminated by additional bacteria.

- Inoculation: After sterilizing the substrate, mycelium or spores from mushrooms are injected into it. Spore solution can be injected or colonized grain or agar can be transferred to the substrate.

2. Embryology

- To promote mycelium growth, the substrate containers are kept in a warm, dark environment after inoculation. Generally, the range of ideal temperatures for incubation is 70°F to 80°F (21°C to 27°C).

• Over a few weeks, the mycelium will invade the substrate and create a dense filament network.

3. Fruiting Environment:

• It's time to encourage fruiting once the substrate has been completely colonized. This entails subjecting the mycelium to circumstances that cause mushrooms to grow.

• Give light: reason why they don't need direct sunshine, magic mushrooms need light to begin fruiting. It is adequate to use natural light filtered through a window or indirect light from fluorescent lights.

- Modify temperature and humidity: Fruiting can be aided by a small drop in temperature (around 65°F to 75°F, or 18°C to 24°C) and an increase in humidity.

- Preserve airflow: The development of mushrooms depends on oxygen exchange and the avoidance of carbon dioxide buildup, both of which are dependent on adequate air circulation.

4. Gathering:

• Once the mushrooms start to grow, they can be harvested just before the spores start to fall, but before the caps have completely opened.

To harvest the mushrooms, carefully twist or cut them at the bottom of the stem. Refrain from harming the substrate or mycelium.

5. After-Harvest Management:

• Before starting a new fruiting cycle, let the substrate rest for a while (referred to as a "rest cycle") after harvesting.

- Get rid of spent substrate properly to keep pests and contaminants at bay.

6. Continue Cycles:

- The cultivation of magic mushrooms is frequently carried out in cycles, involving several rounds of substrate preparation, inoculation, incubation, and fruiting.

- Growers can adjust cultivation practices and environmental factors at the end of each cycle to maximize output and quality.

It is noteworthy that the production of magic mushrooms might be restricted by law in certain areas. Before beginning any growing operations, growers should become acquainted with the laws and regulations that apply in their area. Furthermore, to avoid infection and guarantee successful mushroom production, good sanitation and sterile practices are essential.

Preparation of substrates

In order for mushroom mycelium to colonize and flourish, the preparation of the substrate is an essential step in the development of magic mushrooms.

1.Choosing a Substrate: While magic mushrooms can be grown on a range of substrates, a blend of organic materials, such as grains, straw, or compost, is the most popular and productive type. Vermiculite, brown rice flour, and different grains (such as millet and rye) are common substrates.

2. Sterilization or Pasteurization: The substrate needs to be sterilized or pasteurized in order to get rid of any competing microbes that might prevent the growth of mushrooms before being inoculated with spores or mycelium. Heating the substrate to high temperatures (usually above 100°C) is known as sterilization, and it is used to destroy bacteria, fungi, and other pollutants. Pasteurization is the process of heating a substrate to lower temperatures (around 70–80°C) for an extended amount of time in order to preserve helpful bacteria and kill dangerous organisms on a selective basis.

3. Making the Substrate Mixture: The substrate components are combined to produce a homogenous, moisture-retentive mixture after they have been pasteurized or sterilized. The particular growth technique and mushroom species being grown may have an impact on the substrate mixture's precise composition.

4. Adjustment of Moisture: The mixture of substrates should be damp but not drenched. As necessary, add more water or other moisture sources to the substrate to change its moisture content. When squeezed, the substrate should stay together and not leak any water.

5. Filling Bags or Containers: Next, the substrate mixture that has been made is put into bags or containers that are meant to be used for growing mushrooms. Glass jars, plastic trays, and plastic bags are examples of common containers. Make sure there are no pollutants and that the containers are clean.

6. Sealing and Sterilization: To permit gas exchange while avoiding contamination, containers or bags containing substrate are sealed using the proper closures (such as filter patches, lids). If filling a sealed container, it might be necessary to sterilize it once again to make sure that no contaminants are introduced.

7. Cooling and Inoculation: The substrate needs to reach room temperature after pasteurization or sterilization before being inoculated. To avoid contamination, the substrate is injected with mycelium or mushroom spores using sterile methods once it has cooled.

8. Incubation: To promote the growth of mushroom mycelial cells, inoculated substrate containers or bags are placed in an environment that is warm, dark, and humid. The mycelium colonizes the substrate and forms a dense network of hyphae during this incubation phase.

A key component in growing magic mushrooms is preparing the substrate, which creates the conditions for a successful mycelial colonization and subsequent fruiting. To reduce the chance of contamination and increase mushroom harvests, it is imperative to pay close attention to detail, maintain cleanliness, and use the right sterilizing procedures.

Techniques for vaccination

The procedure of placing mushroom spores or mycelium onto a substrate that has been prepared to start the growth of mushroom mycelium is known as inoculation. There are a few different ways to inoculate substrate with magic mushroom mycelium or spores, and each has pros and downsides

1. Syringe Inoculation for Spores:

One of the easiest and most widely used techniques for introducing magic mushroom spores into a substrate is the Spore Syringe Inoculation method.

- Spores are injected directly into the substrate using a spore syringe that holds a suspension of spores in a sterile liquid solution (often distilled water or a saline solution).

- A tiny amount of spore solution is injected into the substrate by inserting the syringe through an injection port or self-healing injection port.

- Spore syringe inoculation is very easy to do and needs little equipment, which makes it perfect for novices.

2. To inoculate with liquid culture:

- Liquid culture is the process of introducing mushroom spores or mycelium into a liquid nutrition solution and letting the mycelium develop and multiply there.

- By incorporating some of the liquid culture into the substrate, once it has been established and is developing, it can be utilized to inoculate substrate.

- When it comes to producing huge amounts of mycelium for inoculation, liquid culture inoculation may be more effective than spore syringe injection.

3. Agar Immunization:

- Agar inoculation is the process of cultivating mushroom mycelium in a Petri dish or other sterile container using an agar medium that is rich in nutrients.

- After the mycelium has colonized the agar, the mycelium-containing tiny portions (agar wedges) are sliced from the plate and placed on the substrate.

- Agar inoculation can be used to generate a clean, healthy inoculum for substrate inoculation, as well as to select and isolate strong mycelial strains.

4. Inoculation of grain sparrow:

- Using mycelium or spores from mushrooms, sterilized grains (such as rye or millet) are inoculated.

- The grains can be utilized as inoculum to inoculate bigger batches of substrate after they have been completely colonized by the mycelium.

- Grain spawn inoculation can be expanded for commercial output and is frequently utilized in larger-scale mushroom growth operations.

5. Inoculation of Tissue Cultures:

- To produce mycelium, tissue culture entails separating tiny fragments of mushroom tissue, such as the stem or cap tissue, and cultivating them on a nutrient medium.

- The mycelium can be used to inoculate substrate in a way akin to liquid culture or agar inoculation once it has grown and multiplied.

Chapter Five

The growth-promoting environment

Magic mushrooms require the correct environmental conditions to grow and develop successfully. These should be similar to the natural environment that mushrooms live in. These are the main environmental elements to take into account:

1. Temperature: During the colonization phase, when mycelial growth occurs, magic mushrooms typically grow best in temperatures between 70°F and 75°F (21°C and 24°C). During the fruiting phase, when mushrooms form, these temperatures are slightly lower, typically between 65°F and 70°F (18°C and 21°C). It is crucial to keep the temperature steady within these ranges in order to support fruiting and healthy mycelial growth.

2. Humidity: Magic mushrooms need high humidity levels to flourish, particularly during the fruiting phase when they need a lot of moisture to fully develop and mature. While somewhat lower relative humidity values (between 80% and 90%) are appropriate for mycelial growth, levels of 90% to 95% are optimum for fruiting. Techniques like misting, humidifiers, or establishing a humid microclimate inside the cultivation area can all be used to preserve humidity.

3. Light: The process of photosynthesis in magic mushrooms does not require light, in contrast to plants. However, light is important for controlling the fruiting process and determining how mushrooms develop. For mushrooms to fruit, indirect or ambient light is usually sufficient, and the development of mushroom fruiting bodies is triggered by light exposure. In order to promote mycelial growth, mushrooms are typically kept in low light or complete darkness during the colonization period.

4. Air Exchange: In order to remove carbon dioxide and metabolic byproducts from the growing mycelium and to supply it with fresh oxygen, proper air exchange is necessary. Sufficient air exchange also aids in controlling humidity levels and preventing the accumulation of impurities. To attain the best possible air exchange inside the cultivation area, passive or active ventilation equipment, such as fans, air pumps, or natural airflow, can be employed.

5. Substrate Moisture: Magic mushroom growth and development depend on the substrate's proper moisture content being maintained. It is important to keep the substrate damp but not soggy because too much moisture might cause contamination and stunt the growth of mushrooms. To maintain constant moisture during the growing phase, it might be required to periodically mist or water the substrate in addition to routinely checking its moisture content.

Cultivators can establish an ideal growing environment for magic mushrooms that supports vigorous fruiting and healthy mycelial growth by carefully regulating certain environmental parameters. To ensure successful mushroom growth and to suit the specific requirements of the species being farmed, it is important to periodically evaluate and adapt the ambient conditions.

Chapter six

Key psychoactive alkaloids (such as psilocybin and psilocin)

Magic mushrooms contain chemical molecules called psychoactive alkaloids, which give them their hallucinatory properties. Psilocybin and psilocin are the two main psychoactive alkaloids found in magic mushrooms, while additional substances may possibly play a role in their psychoactive effects. Below is a summary of these important alkaloids:

1. Psilocybin:

The main hallucinogenic ingredient in magic mushrooms is psilocybin, which is a member of the tryptamine class of chemicals.

● Psilocybin is not psychotropic in its natural condition. But when consumed, the body quickly converts it into psilocin, the key ingredient that gives magic mushrooms their hallucinogenic properties.

● In the brain, psilocybin partially agonistically interacts with serotonin receptors, specifically the serotonin 2A receptor (5-HT2A). These receptors are activated, which results in changes to mood, consciousness, and sensory perception. These changes can include euphoria, hallucinations, and altered states of consciousness.

2. Psilocybin:

- The principal active metabolite of psilocybin, psilocin, is what causes the immediate effects that are felt after consuming magic mushrooms.

- Psilocin functions as a partial agonist at brain serotonin receptors, specifically the serotonin 2A receptor (5-HT2A), much like psilocybin does. Its effects resemble those of psilocybin in that they involve mood and cognitive alterations, hallucinations, and altered perception.

Since psilocin is the direct active ingredient that causes the psychedelic effects of magic mushrooms, it is more potent than psilocybin in terms of weight.

3. Baeocystin:

• Baeocystin is a less well-known hallucinogenic substance that can be found in some magic mushroom species, albeit in less concentrations than psilocybin and psilocin.

- Baeocystin is thought to contribute to the overall psychedelic experience caused by magic mushrooms, sharing structural similarities with psilocybin and psilocin. Its precise effects and modes of action, however, remain little understood.

4. Norbaeocystin:

- Another small hallucinogenic ingredient, norbaeocystin, is present in some species of magic mushrooms but usually in smaller amounts than psilocybin and psilocin.

Norbaeocystin shares structural similarities with psilocybin and psilocin, much like baeocystin, and it could potentially add to the overall psychedelic effects of magic mushrooms. Its precise effects and modes of action, however, have not been thoroughly investigated.

The main psychoactive alkaloids in magic mushrooms cause changes in perception, emotion, and awareness that are typical of a psychedelic experience by interacting with serotonin receptors in the brain, specifically the serotonin 2A receptor. To completely comprehend the effects of these substances and their possible therapeutic uses, more investigation is required.

Chapter Seven

These substances' effects on the human brain

Psilocybin and psilocin, two psychoactive substances present in magic mushrooms, have a variety of intricate effects on the human brain. These substances mostly affect the brain's serotonin receptors, specifically the serotonin 2A receptor (5-HT2A), which changes brain activity and neurotransmitter communication. The effects of psilocybin and psilocin on the human brain are summarized as follows:

1. Modified Perception and Sensory Integration:

- Seizures such as visual, auditory, and tactile perception can all be profoundly altered by psilocybin and psilocin. Intense sensory experiences, improved color perception, and complex and vivid visual hallucinations are possible for users.

- It is thought that these changes in perception are caused by modifications in the activity of brain regions involved in sensory processing, such as the visual and auditory cortex, as well as changes in how sensory information is integrated throughout brain networks.

2. Mood and emotional swings:

• A wide spectrum of emotional states, such as euphoria, introspection, and emotional release, can be induced by magic mushrooms. Emotional openness, empathy, and connectivity are among the emotions that users may report.

• The modification of serotonin transmission in brain areas involved in emotion regulation, such as the amygdala and prefrontal cortex, is hypothesized to mediate these changes in mood and emotion. While decreasing activity in brain regions linked to negative emotions like fear and anxiety, psilocybin and psilocin may increase activity in circuits linked to good mood states.

3. Ego disintegration and altered awareness:

• Two of the most important effects of magic mushrooms are the alteration of consciousness and the sensation of ego dissolution, which is characterized by a temporary loss of sense of self and identity.

• Psilocybin and psilocin cause disruptions to the default mode network (DMN), a network of brain regions involved in introspection and self-referential cognition. This disruption leads to a feeling of connectedness to the outside world and one's surroundings, which lowers one's ego.

4. Improved Neural Connectivity and Synchronization:

- Neuroimaging research has shown that psilocybin and psilocin improve neuronal synchronization and connectivity in the brain, particularly between regions linked to higher-order cognition and sensory processing.
- These changes in brain connections are thought to be the

origin of the altered states of consciousness and sensory experiences brought on by magic mushrooms, as well as the increased creativity and insight reported by some users.

5. Positive Outcomes:
- Beyond their immediate effects on perception and consciousness, magic mushrooms have shown promise as therapeutic agents for treating a range of mental health disorders, including addiction, PTSD, anxiety, and depression.
- The therapeutic effects of psilocybin and psilocin are thought to be attributed to their ability to promote neuroplasticity,

enhance emotional processing, and induce profound shifts in viewpoint and worldview.

The effects of psilocybin and psilocin on the human brain are complex and varied, involving modifications to perception, emotion, consciousness, and neuronal activity. Further research is necessary to fully understand the processes behind these benefits and their potential therapeutic applications.

Pharmacological features and metabolism

The primary psychedelic components found in magic mushrooms, psilocybin and psilocin, differ in their pharmacological properties and rates of metabolism. These variations play a significant role in the effects that these substances have on the human body and brain. Below is a summary of their pharmacological and metabolic properties:

Pharmacology's characteristics:
1. Agonism of Serotonin Receptors: Psilocybin and its active metabolite psilocin primarily function by partially agonistically attaching to brain serotonin receptors, particularly the serotonin 2A receptor (5-

HT2A). The activation of these receptors results in modifications to neurotransmitter signaling and brain activity, which in turn causes the hallucinogenic effects of magic mushrooms.

2. Neurotransmitter Modulation: In addition to serotonin receptors, psilocybin and psilocin interact with the brain's dopamine, glutamate, and norepinephrine systems. These interactions affect the complex effects that magic mushrooms have on emotion, perception, and cognition.

3. Disruption of the Default Mode Network (DMN): The brain networks involved in introspection and self-referential thought are part of the default mode network (DMN), which is disrupted by psilocybin and psilocin. This disruption causes a temporary loss of identity and dissolution of the ego, which is characteristic of the hallucinogenic experience caused by magic mushrooms.

4. According to recent research, psilocybin and psilocin may promote neuroplasticity and synaptic remodeling in the brain, which may lead to long-lasting changes in neural connections and behavior. These neuroplastic actions may explain the magic mushroom's medicinal promise for treating a variety of mental health issues.

The process of metabolism

1. Conversion to Psilocin: Upon ingestion, psilocybin is rapidly transformed by the body into psilocin, its active form. Alkaline phosphatase and other liver enzymes are primarily responsible for this process. Psilocin is the chemical that gives magic mushroom users their initial affects.

2. Distribution: After it has formed, psilocin is widely distributed throughout the body and can readily cross the blood-brain barrier to influence the central nervous system. One to two hours after ingestion is when peak blood psilocin concentrations often occur, though this can vary depending on a number of factors, such as dosage, metabolism, and whether or not food is in the stomach.

3. Elimination: Psilocin is further metabolized in the liver by enzymes such monoamine oxidase (MAO) and aldehyde dehydrogenase (ALDH), producing inactive metabolites that the body eliminates in urine. The half-life of psilocin elimination is brief, typically lasting two to four hours.

4. Individual Variability: The metabolism of psilocybin and psilocin can be affected by a variety of factors, including genetics, liver function, drug usage, and other substances that change enzyme activity. Variations in metabolism can affect the onset, duration, and intensity of psychedelic effects after consuming magic mushrooms.

Understanding the pharmacological properties and metabolism of psilocybin and psilocin is necessary to clarify their impact on the human body and brain as well as potential therapeutic applications. Further research is needed to fully understand the mechanisms underlying these compounds and how they interact with the body and brain.

Chapter Eight

Techniques for Extracting

The methods used to extract psilocybin and psilocin from magic mushrooms varies in complexity and efficacy. These methods often include extracting the psychoactive compounds from the mushroom material using solvents, then purifying the extracted compounds to distinguish the desired ones. A few common extraction methods are as follows:

1. Easy extraction with solvent

• The hallucinogenic ingredients in dried magic mushrooms are extracted using a solvent such as ethanol, methanol, or acetone. After that, the mushrooms are finely powdered.

• The powdered mushroom material is steeped in the solvent for a predetermined period of time to extract psilocybin and psilocin.

• After the solvent has been filtered to eliminate solid particles, the extracted extract is evaporated to concentrate the psychotropic components.

2. Extraction of Base-Acid:

• The more advanced method of separating psilocybin and psilocin from magic mushrooms is known as acid-base extraction, and it involves a number of steps.

• The powdered mushroom material is first soaked in an acidic solution (such as hydrochloric acid or citric acid) to convert psilocybin into its salt form that is soluble in water.

• The acidic solution is then basified, its pH is raised, and psilocin is transformed into its freebase form, which is soluble in non-polar solvents, by adding a strong base, such as sodium hydroxide.

• Next, a non-polar solvent (such petroleum ether or diethyl ether) is added to the mixture in order to extract the psilocin from it.

• Following extraction, the solvent is discarded to yield a crude extract containing psilocybin and psilocin. The extract can then be purified even further via chromatography or recrystallization.

3. Chromatographic Techniques:

• Chromatographic techniques like thin-layer chromatography (TLC) and column chromatography can be used to separate and purify psilocybin and psilocin from other ingredients in the extract. These methods, which are based on the differential partitioning of compounds between a stationary phase and a mobile phase, enable the separation of components based on their chemical properties.

- After chromatographic separation, the desired compounds can be removed from the column or TLC plate and collected for further analysis or purification.

4. Recrystallization

- Recrystallization is a purification technique in which a solvent-dissolved crude extract containing psilocybin and psilocin is allowed to progressively cool to allow the necessary components to separate out.

- The resulting crystals can be cleaned with the proper solvent after being centrifuged or filtered to remove impurities.

• Recrystallization can yield very pure psilocybin and psilocin suitable for analytical or medicinal use.

These are only a few examples of the methods used in the extraction process to isolate psilocybin and psilocin from magic mushrooms.

Chapter Nine

Traditional techniques for extracting

For decades, indigenous communities all over the world have been employing ancient methods to extract hallucinogenic chemicals from magic mushrooms. These techniques may vary depending on cultural norms and the resources available, but they frequently ask for simple steps and locally sourced materials. Here are some common extraction techniques:

1. Tea Brewing
One simple traditional extraction

method is to soak magic mushrooms in tea. Fresh or dried mushrooms are boiled in water for an extended period of time to dissolve the chemicals that give them their intoxicating properties.

• The mushroom tea is then consumed orally, and the advantages typically manifest themselves within 20 to 30 minutes and continue for several hours.

• This method is favored because it is uncomplicated, easy to make, and reduces the gastrointestinal distress associated with consuming raw mushrooms.

2. Ingestion or Chewing:

• Many aboriginal societies consume magic mushrooms orally, either by chewing the dried or fresh mushrooms or by ingesting them whole.

• The effects of the substance are similar to those of brewed tea when chewed or consumed because it allows the psychoactive ingredients to enter the body through the digestive system and be digested.

• This strategy may be preferred in cultures where brewing tea is customary or where consuming mushrooms is a part of religious or spiritual rites.

3. Preparing it as food:

- In some cultures, magic mushrooms can be added, either raw or cooked, to traditional cuisines or recipes. For example, you can add mushrooms to soups, stews, and other foods.

- Cooking mushrooms may alter their bioavailability and affect the timing and duration of effects in comparison to consuming them raw.

- This method makes it possible to share meals and experiences with people while also enabling the tasty and culturally acceptable ingestion of mushrooms.

4. Sun-drying or smoking

- Before being dried in the sun or by fire, magic mushrooms are sometimes smoked or vaporized in certain cultures.

- Smoking or vaping mushrooms speeds up the absorption of psychoactive chemicals through the lungs compared to oral ingestion, resulting in effects that are felt more quickly.

• This method may be used in indigenous civilizations in place of oral ingestion or as a part of specific rituals or ceremonies.

These traditional extraction techniques demonstrate how creatively and resourcefully indigenous peoples have used the hallucinogenic properties of magic mushrooms for therapeutic, spiritual, and cultural purposes. Even if modern procedures have greatly improved, traditional extraction methods are still valued for their cultural significance, practicality, and ancestral wisdom.

Chapter Ten

Extraction techniques employed in labs

Modern laboratory approaches for extracting and isolating psychotropic compounds from magic mushrooms require complex techniques and equipment in order to attain high purity and efficiency.

Below is a summary of the methods used in laboratories today to extract psilocybin and psilocin:

1. Getting the Sample Ready:

• Magic mushrooms are typically dried and ground into a fine powder in order to increase surface area and facilitate extraction.

• Sample homogenization is carefully carried out to ensure consistency and repeatability in the subsequent extraction procedures.

2. Extraction of solvent

• The powdered mushroom material is extracted, and any hallucinogenic components, such as psilocybin and psilocin, are dissolved, using an appropriate solvent.

• Because acetone, methanol, and ethanol are extremely soluble in psilocybin and psilocin, they are commonly used as extraction solvents.

- Maceration, ultrasonic extraction, or Soxhlet extraction are some techniques that can be used throughout the extraction process to improve production and efficiency.

3. Filtration and Concentration:

- Filtering the extract to remove impurities and solid particles yields a crude extract that contains psilocybin and psilocin.
- The solvent is then evaporated under low pressure or by using techniques like rotary evaporation or freeze-drying in order to concentrate the extract and remove any leftover solvent.

4. Purification:

• The crude extract is refined in order to isolate psilocybin and psilocin from other components of the extract.

• Chromatographic techniques such as preparative liquid chromatography and column chromatography are widely used in purification processes. Thanks to these techniques, it is possible to separate psilocybin and psilocin based on the chemical properties of the respective compounds.
Thin-layer chromatography (TLC) can also be used to first separate

and identify target molecules before further purification.

5. Formation of Crystals:

• Recrystallization is a process that involves dissolving psilocybin and psilocin in a suitable solvent and allowing the compounds to crystallize into crystalline crystals in order to further purify them.

• The compounds undergo a procedure called crystallization that enhances their purity and removes any remaining impurities, resulting in psilocybin and psilocin that are very pure and suitable for use in medical research.

6. Assessment and Quality Control:

• To verify the identification and purity of separated psilocybin and psilocin, analytical techniques such as nuclear magnetic resonance (NMR) spectroscopy, gas chromatography-mass spectrometry (GC-MS), and high-performance liquid chromatography (HPLC) are utilized.

• Quality control methods ensure that the final product meets safety, potency, and purity standards established by the analytical or pharmaceutical sectors.

These modern laboratory techniques enable the efficient and precise isolation and purification of psilocybin and psilocin, opening up new research directions and opportunities for medication development and other applications.

Chapter eleven

Purification and Isolation of Psilocybin

Psilocybin, the primary hallucinogenic component of magic mushrooms, is extracted and separated in contemporary labs by a combination of extraction, chromatography, and crystallization techniques.

1. Extraction

• The first step involves grinding dried magic mushrooms into a fine powder in order to maximize the surface area available for extraction.

- Next, the powdered mushroom material is extracted using a suitable solvent, such as ethanol or methanol. Dissolving psilocybin and other target compounds while dissolving contaminants and insoluble plant materials is the aim of this step
- Psilocybin extraction techniques such as maceration, reflux, or Soxhlet extraction can be employed to ensure a successful result.

2. Filtration and Concentration:

- Following extraction, the solvent is filtered to remove debris and solid particles. Following filtration, the extract can be concentrated, and excess solvent can be eliminated by rotational evaporation or evaporation, resulting in a crude extract that includes additional compounds including psilocybin.

Because psilocybin can decay under certain conditions, it's vital to limit exposure to light and oxygen during this procedure.

3. Removal of Chromatic Contamination:

• Chromatography is widely used to separate and purify psilocybin from other ingredients in the crude extract.

• A common chromatographic technique for this is column chromatography, in which the crude extract is placed onto a column filled with a stationary phase (such silica gel or reverse-phase resin).

• As the solvent (mobile phase) moves through the column, the different extract elements are kept and separated selectively based on how well they bind to the stationary phase. Psilocybin can then be eluted from the column using the proper solvent solution.

• Alternatively, more accurate and efficient methods for purifying psilocybin include preparative thin-layer chromatography (TLC) and high-performance liquid chromatography (HPLC).

4. Crystallization:

• Following chromatographic purification, the psilocybin-containing fractions are concentrated and crystallized to further purify the drug.
• Inducing the crystallization process can involve adding the proper antisolvent or gradually removing the solvent under closely supervised conditions.
• The resulting psilocybin crystals are collected by filtration or centrifugation, decontaminated with a cold solvent, then vacuum-

dried to get pure psilocybin powder.

5. Explanation and Analysis:

- The pure psilocybin product is analyzed and characterized using nuclear magnetic resonance (NMR) spectroscopy, mass spectrometry, and high-performance liquid chromatography (HPLC).

- You may be certain that the psilocybin product is authentic, pure, and meets the requirements necessary for research or medical use by employing these analytical techniques.

Potential benefits for depression, anxiety, addiction, PTSD, etc.

Psilocybin, the primary hallucinogenic component of magic mushrooms, has demonstrated potential as a treatment for a variety of mental health conditions, including as addiction, PTSD, depression, and anxiety. Psilocybin-assisted therapy appears to provide the potential benefits listed below, based on preliminary studies in this area

1. Depression

• Clinical trials have shown that psilocybin-assisted therapy can rapidly and sustainably alleviate symptoms of treatment-resistant depression.

• Psilocybin appears to improve mood, self-awareness, and overall health by promoting neuroplasticity and enhancing emotional processing.

• Studies show that a single dose of psilocybin combined with psychotherapy can provide significant and long-lasting antidepressant effects; some patients experience extended periods of time without experiencing depressive symptoms after treatment.

2. Anxiety Disorders:

• Psilocybin-assisted therapy has demonstrated promise in the treatment of anxiety disorders, including generalized anxiety disorder (GAD), social anxiety disorder (SAD), and existential anxiety.

• Psilocybin may be able to help people confront and reframe existential problems and worries, which might lessen anxiety and fear of death, because it can induce mystical or transcendent experiences.

• Studies have demonstrated that psilocybin-assisted therapy greatly lowers anxiety and improves quality of life; these effects can last for several months to years following treatment.

3. Post-traumatic stress disorder, or PTSD

Preliminary study suggests that psilocybin-assisted therapy could be helpful in improving overall health and reducing symptoms of post-traumatic stress disorder.

• Individuals with PTSD may find it simpler to confront and incorporate traumatic memories and experiences if psilocybin facilitates emotional processing and therapeutic insights. Positive initial results imply that psilocybin-assisted PTSD treatment lessens symptoms and improves quality of life. Currently, clinical trials are being carried out to evaluate the efficacy and safety of this medication.

4. Addiction-related disorders:

• Studies have shown that opiate, nicotine, and alcohol addiction are among the drug use disorders that may be treated with psilocybin-assisted therapy.

• People may be able to overcome rooted addiction and dependent habits because psilocybin has the ability to evoke profound mystical experiences and insights.

Following psilocybin-assisted therapy, clinical trials have shown improvements in mood, self-efficacy, and quality of life in addition to decreases in cravings, withdrawal symptoms, and relapse rates.

5. Existential distress and fear of death:

• Patients with cancer and other life-threatening illnesses who are feeling existential distress and end-of-life fear may benefit from psilocybin-assisted therapy.

• Psilocybin may help people face and overcome existential worries and anxiety related to death since it can induce mystical or transcendent experiences.

- Clinical trials have demonstrated that psilocybin-assisted therapy greatly reduces existential distress, despair, and anxiety in individuals with life-threatening conditions. Furthermore, the general well-being and quality of life of the patients have improved.

Chapter Twelve

Examining spirituality and consciousness

Psilocybin, the hallucinogenic component of magic mushrooms, has long been prized for its capacity to promote conscious and spiritual growth. Native American tribes and spiritual seekers have been using magic mushrooms for eons as sacraments in religious ceremonies, shamanic practices, and visionary experiences. In spiritual and conscious exploration, psilocybin can help in the following ways:

1. Mysterious and Unreal Experiences:

- Psilocybin's amazing capacity to induce mystical or transcendent experiences characterized by an ego collapse and a feeling of connection and interconnectedness.

- During these experiences, people may report having strong feelings of awe, reverence, and closeness to the cosmos, a higher power, or a group consciousness.

- Mysticism brought on by psilocybin can result in spiritual insights, existential reflection, and a profound sense of life's meaning and purpose.

2. Enhanced Awareness and Sensation:

• By enhancing sensory awareness and perception, psilocybin can alter tactile, auditory, and visual perception.

• Individuals might notice sharper perceptions, more vibrant colors and patterns, and a greater appreciation for the complex beauty of the natural world.

• These shifts in perception have the potential to promote a stronger sense of present, mindfulness, and connectedness with one's surroundings and inner experiences.

3. The Self and Ego Are Transcendent:

- The ability of psilocybin to momentarily dissolve an individual's ego leads to a profound sense of unity with other individuals, the environment, and the universe.

People who transcend their sense of self and become a part of a bigger whole might develop feelings of humility, compassion, and empathy through experiences of ego dissolution.

This dissolution of the ego can lead to a deeper sense of connectedness with all living things, spiritual development and self-discovery.

4. Symbolic and archetypal images:

• The unconscious mind is usually the source of the vivid and symbolic imagery linked to psilocybin experiences, which includes spiritual symbolism, mythical themes, and archetypal patterns.

- People may come upon visionary sceneries, magical creatures, and sacred geometry. More in-depth reflection, symbolic integration, and spiritual growth can all be aided by these encounters.

- Symbolic imagery seen during psilocybin experiences may resonate with themes of death and rebirth, oneness and dualism, and the interconnection of all things.

5. Assimilation and Reconciliation:

- Psilocybin experiences have the potential to initiate profound healing and transformation on a psychological, emotional, and spiritual level.

- By confronting and resolving unresolved traumas, fears, and limiting beliefs, people can experience tremendous healing, forgiveness, and acceptance. Integrating psilocybin involves thinking about and bringing fresh viewpoints into daily life, relationships, and spiritual pursuits.

6. The connection between nature and culture:

Psilocybin experiences have the potential to evoke a profound sense of gratitude towards the natural environment and indigenous wisdom traditions that have traditionally held magic mushrooms in high esteem.

• There may be a renewed commitment among people to environmental preservation, social justice, and the preservation of indigenous cultures and sacred sites.

- Psilocybin fosters an understanding of ecological consciousness and oneness with all living species, which may improve appreciation for the interwoven web of life on Earth.

Psilocybin generally has the potential to facilitate profound spiritual and consciousness exploration, offering opportunities for individuals to grow personally, discover more about themselves, and become increasingly connected to life's mysteries. People can be deeply changed, inspired, and enriched by these interactions, which can also help them develop compassion, self-awareness, and harmony with the outer world.

Conclusion

In conclusion, psilocybin, the hallucinogenic component of magic mushrooms, exhibits significant potential for a range of applications, such as therapeutic interventions and spiritual investigation.

Remarkable therapeutic efficacy has been shown by psilocybin-assisted treatment in treating a variety of mental health issues, such as addiction, PTSD, anxiety, and depression. Clinical research has demonstrated that it can result in notable and enduring modifications in behavior, mood, and perception. This offers new hope to those who have not responded to conventional therapy.

Beyond its medicinal use, psilocybin is incredibly valuable for spiritual and consciousness exploration. Its power to elicit mystical experiences, enhance consciousness, dissolve the ego, and foster a sense of oneness with all living forms has been revered for ages by indigenous civilizations and spiritual practitioners. Psilocybin experiences can lead to profound insights, healing, and metamorphosis that broaden one's perspective of the universe, oneself, and the intricate web of life.

Since the body of knowledge on psilocybin is still expanding, it's critical to utilize it sensibly, respectfully, and with care. Regarding the potential advantages and disadvantages of psilocybin for both personal and societal healing and development, it is imperative that we safeguard people's health, advance ethical and sustainable behaviors, and honor indigenous wisdom traditions.

Through the application of psilocybin's therapeutic and

spiritual capabilities, we can promote greater peace, empathy, and compassion in the environment as well as inside ourselves. With further research, education, and integration, psilocybin may offer important insights and avenues toward healing, transformation, and communal growth.

www.ingramcontent.com/pod-product-compliance
Lightning Source LLC
Chambersburg PA
CBHW071058240526
45471CB00016B/2151